Strange Scenes In Death Valley

Climate Chaos In The California Desert

Jan Stewart

Copyright © 2024 | Jan Stewart

All rights reserved.

No part of this publication may be reproduced, distributed, or transmitted in any form or by any means, including photocopying, recording, or other electronic or mechanical methods, without the prior written permission of the publisher, except in the case of brief quotations embodied in critical reviews and certain other noncommercial uses permitted by copyright law.

TABLE OF CONTENTS

Introduction

Chapter 1: A Roller Coaster in the Parched Earth of Death Valley

Chapter 2: A Soaked Desert: The Opening of the Heavens

Chapter 3: A Desert in Bloom: Death Valley's Technicolor Revival of Nature

Chapter 4: In an Extreme Future, Can Death Valley Withstand the Whiplash?

Introduction

Over the past few years, Death Valley has provided some of the strangest sights in the boom-and-bust environment of California.

A protracted drought that impacted the region until 2022 caused some of the perennial creosote plants in the area to die back. Then, heavy rains from Hurricane Hilary's aftermath and later storms brought back annual wildflowers that had been dormant all previous year.

An ancient lake that washed out throughout the winter is now starting to drain again thanks to heavy rainfall.

These extremes together have created strange juxtapositions in the well-known desert.

"I could lead you to an area of dead creosote bushes with lovely wildflowers emerging between," stated

Patrick Donnelly, a Center for Biological Diversity conservation biologist. According to plants, the area is a wasteland after the end of the world where everything is dead. However, as spring arrives, flowers begin to appear among the corpses.

Climate scientists have long predicted that rising temperatures will intensify rainfall events and cause more droughts in California's interior desert regions. Locals in Death Valley claim to be seeing that dynamic unfold in real time. The future of this emblem of resilience will depend on how local economies, plants, and animals in the desert adjust to changes in an already extremely hard environment.

"Having a front row seat is interesting," Donnelly continued. "This is how climate chaos is identified."

Twice in a 1,000-year storm
The local climate, according to Susan Sorrells, proprietor of the ecotourism village of Shoshone,

which is close to Death Valley National Park, is "always a roller coaster ride."

But it has been especially true during the last few years. Beginning in early 2022, Death Valley, along with the rest of the American Southwest, was submerged in the driest spell to hit the region since the year 800. The 22-year drought resulted in the lowest soil moisture levels ever measured. Plants were wilting due to the deep roots of the creosote bush, which has been known to endure for thousands of years and can withstand very little.

Its continued growth during the drought was evidence of how bad the weather was.

According to University of California Riverside research ecologist Lynn Sweet, desert plants are extremely hardy and die slowly. "But mortality sets in at some point."

Then, in August of that same year, there was rain.

In a few hours, Death Valley National Park received an unprecedented 1.7 inches of rain, or nearly three-quarters of the park's annual total. Meteorologists labeled the storm a once-in-a-1,000-year occurrence that wiped away park roads, moved rocks, and trapped cars in debris.

After a year or two, the remnants of cyclone Hilary, an uncommon Pacific storm that smashed all prior records for 24-hour rainfall, dumped 2.2 inches of rain on the park in a single day.

"It's obvious that climate whiplash is to blame for having both the greatest drought and the most precipitation in recorded history," said Donnelly.

The storm also shut down Highway 127, which connects the park to gateway villages, converting it into "a raging river," according to Sorrells.

She operates an ecotourism business that includes a 20-bed inn, hiking trails, a trading post, an RV park, and the Crowbar Cafe and Saloon. Sorrells was in a dire situation because the highway repair was supposed to last through the winter and into spring, making the area almost impossible to access during the peak season.

She stated that there was a joke in the area that said we made all of our money from October through the end of April, and then we lost it from May through September.

Sorrells claims that the locals banded together and put pressure on state representatives to get the project along more quickly. The park partially reopened in October, and the road reopened in January.

Then something interesting happened.

Donnelly claims that spring wildflowers that sprouted in the fall endured a warm winter, despite the fact that cold weather would ordinarily cause them to wither back down. And in February, a multiday atmospheric river storm event caused another flood.

"The atmospheric river was able to replenish the wildflowers with all this rain," Donnelly said. They are currently growing in an unusual manner. These wildflowers have sturdier stems, are much higher than usual, and are incredibly powerful.

And there was enough water to revive the ancient, usually dust-covered lakebed of Lake Manly.

NASA reports that when the National Park Service first announced in February that kayakers would be able to paddle on the lake, it was as deep as three feet.

Sorrells said that "people really came out for the sensational and fell in love with the desert's beauty, charm, and uniqueness." "We just started growing so quickly that we nearly lost control of the company."

dry conditions with sudden flooding
Most climate models indicate that as California's deserts get drier over time, the atmosphere's improved ability to hold on to energy and water will cause occasional storms to get stronger as well.

Sweet said, "Longer droughts, flashier storms."

The past several years have been compared by some to a test run for a new reality.

Many desert species are either inactive or hibernate as a result of drought, which can give the rainy years the appearance of being a relative pageant of color and activity.

For instance, to withstand dry seasons, Mojave Desert tortoises dig underground burrows. Female tortoises can retain their sperm for years, waiting for the right weather to fertilize their eggs.

Some hard-coated desert wildflower seeds can live for hundreds of years until the right conditions are met.

Because Death Valley is so harsh, it is vital to assume that the creatures and plants that live there have developed to the extent that they can survive there. This gets harder as the temperature rises, according to Sweet.

Now the rain is bringing new life.

New bushes are beginning to sprout, but Sweet says they'll need a "nice series of wet years" to flourish. Wildflowers feed lizards, which in turn feed coyotes and other animals, as well as harvester ants.

Sweet expressed optimism that ultimately more resources will find their way down the food chain.

Future temperatures are expected to rise and get drier, and she expressed optimism that "Hopefully we'll get these recovery periods of precipitation to refill the seed bank and energy reservoir of the whole ecosystem."

Chapter 1: A Roller Coaster in the Parched Earth of Death Valley

The phrase "Death Valley" conjures up images of a desolate area where the sun beats down on cracked rock. But over the past three years, this well-known desert has experienced an amazing rollercoaster of weather extremes that has left scientists baffled and the locals feeling both fearful and in awe.

The normally mottled brown and beige earth turns a powdery gray as the greatest drought in recorded history tightens its grip. Picture yourself staring at twisted creosote bushes, some of which are older than your great-great-grandparents. This was no ordinary dry period; rather, it was a parched party crasher that could have permanently turned Death Valley into a dust bowl.

Who is to blame? Unbroken years of drought not seen since the year 800. As each season passed, the nation, accustomed to minimal precipitation, grew increasingly arid. Even the hardiest plants, like the creosote bush with its vast taproots, started to show signs of strain. Normally a dusty green, their leaves took on an unnerving yellow tint, accompanied by a silent scream for water that echoed through the vast emptiness.

The slow-moving desert tortoise, which could spend months underground to avoid the heat, and other creatures that had once lived in Death Valley faced a new threat: their regular food supplies ran out, and even in their clay shelters, the intense sun threatened to roast them to death. This was an indication of a shifting ecology, not just a botanic disaster.

The impact on people was equally dramatic. Susan Sorrells, the owner of an ecotourism company near Death Valley National Park, remembers those days grimly: "It felt like the lifeblood was draining out of the desert." Tourists who usually flocked to Death Valley to see its otherworldly beauty stopped going. The once-vibrant town of Shoshone, which serves as the park's entrance, grew strangely silent. Just as bleak as the seemingly endless stretch of cracked earth seemed, so too did the future of this paradisiacal desert.

But Death Valley, it seems, is not one to give up easily; this was only the start of a play that would alter this unwavering nation's history entirely; it was by no means the end of the saga. Buckle up, because as we turn the page, we are going to witness a shift so significant that it will leave even the most seasoned meteorologist speechless.

The phrase "Death Valley" conjures images of a desolate area where life struggles to survive in a harsh solar environment. But in recent years, this well-known desert has experienced an amazing rollercoaster of weather extremes that has left locals and experts alike perplexed and wondering what in the world—or should we say, beyond Earth—is happening.

The area was dry, and the usually colorful creosote bushes—some of which may survive for thousands of years—were wilting in the scorching sun. Their deep roots, which are often adept at locating water, were simply unable to keep up. It was a spectacle of death in slow motion, a sobering reminder of how fleeting life can be in this harsh world. A few years ago, Death Valley was gripped by a bone-dry drought. The kind of drought when heat waves shimmer in the air and your throat feels like sandpaper.

It was like something out of a post-apocalyptic movie, a terrifying glimpse of what Death Valley would become if the drought continued. Picture yourself driving down a dirt road in a world where the only sound breaking the silence is the sun's unceasing hum. The once-vibrant creosote bushes stood like skeleton sentinels, these twisted, gnarled veterans of the desert, with their foliage a dead, drab brown. It seemed as though even the hardy desert animals had vanished, seeking refuge in the cool embrace of their deep tunnels from the unceasing heat.

But Death Valley is a place of extremes, and it surprised me, as it often does, to see the unthinkable happen in August 2022: the sky parted, and rain pelted the parched floor—brilliant, life-giving showers. Not a little dusting, no, an enormous downpour of biblical proportions. In a few hours, more than 1.7 inches of rain fell—roughly three-quarters of Death Valley's average annual rainfall! It was a once-in-a-thousand-year storm that made meteorologists gape and sigh in retrospect.

Rain fell in torrents, providing much-needed moisture but also wreaking havoc, sweeping away park roads, tossing automobile-sized stones like pebbles and trapping cars in mud rivers; the landscape had changed dramatically from the parched countryside of just a few months earlier.

But this was not the end of the story; a year later, as if determined to test the limits of Death Valley's resilience, another blow was delivered: the remnants of Hurricane Hilary, an unusual Pacific hurricane, dumped an additional 2.2 inches of rain on the park. Talk about whiplash—that amount of rain in a single day is more than Death Valley typically gets in a year! Highway 127, which was vital for connecting Death Valley to other communities, became a raging torrent, forcing the park to close and obstructing access to the region.

Running an ecotourism business near Death Valley, Susan Sorrells remembers the chaos well. Her company depends on a steady stream of visitors, so the situation was dire and the closure of the highway threatened to ruin their busy season. However, like the desert, Susan is resilient; she rallied the community and pressed state representatives to expedite the repairs of the highway, and their efforts paid off when the park reopened in October and the roadway in January.

As you can see, Death Valley's weather has been erratic lately. It's a country of extremes, where floods and droughts can coexist, confusing scientists and locals alike. Nevertheless, Death Valley has shown remarkable resilience throughout, proving that nature can adapt and persevere. This wild ride is far from over, so buckle up. In the next chapter, we'll look at how this unexpected amount of water has altered the desert environment and given rise to an unexpected surge of life.

Chapter 2: A Soaked Desert: The Opening of the Heavens

The phrase "Death Valley" conjures images of a desolate, merciless land. For years, that is exactly what the earth looked like, cracked and parched from the sun's relentless heat. Then, however, something amazing happened: the sky opened, releasing an incredible amount of rain that transformed the desert beyond anyone's wildest expectations.

Travelers narrowed their eyes as they studied the unrelenting sunlight, their cameras capturing the desolate beauty of a parched desert; however, something was stirring beneath the surface, something different promised by the unseen tension that crackled in the air. It all began in August of 2022, after years of drought. The land was already bone dry and appeared to be holding its breath.

And then the storm came, not so much a light sprinkle as a violent downpour that battered the dry ground, each drop a tiny rebellious explosion; the normally placid washes became rushing rivers, creating temporary canyons out of the scorched ground; the earth, unaccustomed to such heavy rain, soaked it up like it was dying.

It poured rain for several hours, and even at Death Valley National Park, where silence is the norm, they heard the sound of water flowing. The rangers, not believing what they were seeing, watched in disbelief as normally dry washes turned into impassable rivers, and cars submerged in the sudden downpour were carried away, their occupants clinging to life as the floodwaters rose.

The normally bone-dry earth glinted with the newly discovered liquid as the storm finally passed, leaving the countryside stunned and drenched. The air was heavy with the scent of petrichor, the earthy aroma typical of the first rain after a protracted dry spell, and carried a strange sense of expectancy.

The storm's effects were immediate and catastrophic: the raging floods flung boulders, normally immovable sentinels of the desert, around like pebbles, and park roads, which had once served as arteries connecting tourists to the heart of the desert, were swept away, leaving behind a jumble of muck and trash.

Shoshone, a small tourist community bordering the park, faced a dilemma when their lifeline to the outside world, Highway 127, turned into a rushing torrent that effectively sealed them off from any potential visitors. Fortunately, the human cost was minimal because park guards, ever vigilant, had evacuated susceptible locations before the storm's full force erupted.

"Spring is when we make the most money," said Susan Sorrells, owner of a successful ecotourism business in Shoshone, as she watched in dismay as the normally steady stream of tourists shrank to a trickle. "Now, with the road closed, the season is practically wiped out."

But in the midst of the devastation, a strange feeling of surprise emerged. Everyone gasped as the floodwaters receded, revealing a scene that had changed. The once barren washes now filled to overflowing capacity with essential water became a haven for desert wildlife, blinking in the unusual moisture as creatures that had sought shelter during the drought surfaced.

When storm Hilary's remnants, an unusual Pacific storm, decided to make landfall in Death Valley a year later, the story of the desert took a new turn. The normally clear sky turned ominous gray as the storm unleashed its fury, battering the already saturated earth with over two inches of rain in a single day—above the park's annual rainfall average—shattering yet another record and demonstrating the desert's growing unpredictability.

The relentless rain had a major impact. Road 127, which had hardly been rebuilt since the previous flood, was once again a raging torrent. Sorrells inspired the neighborhood, and together they convinced state legislators to expedite the vital highway's rehabilitation. The route was finally reopened in January, right in the middle of the busiest travel season, despite years of protest.

But the most intriguing effect of the rain was still hidden until much later: for the first time in its history, the desert began to bloom; spring wildflowers, which usually go dormant in the hot summer months, began to sprout as a result of the unexpected fall precipitation; and it was the warm winter, not the typical lows of below freezing, that caused them to flourish.

Excitement was in the air as conservation biologist Donnelly exclaimed, "These blossoms are giants!"They're taller, thicker-stemmed – like bodybuilders compared to their usual scrawny selves." Tourists were drawn to these surprise blooms, vivid bursts of color against the sandy backdrop of the desert. Sorrells' company took off as people traveled to view the surprising alteration of the desert.

Not only were there fewer wildflowers, but a shimmering pool of water had taken the place of Lake Manly, an old lakebed that was once a vast area of cracked earth.

Chapter 3: A Desert in Bloom: Death Valley's Technicolor Revival of Nature

Death Valley: the name alone conjures images of a desolate place, a furnace under the intense heat of the desert sun. But after the incredible rains of 2022, something happened that even veteran park rangers could not have predicted: this was not the Death Valley you knew. It was a riot of color and life, the rebirth of a desert.

This was not a mirage, but rather a dream come true in the middle of the desert, with golden marigolds stretching toward the light and their happy faces a striking contrast to the worn rocks surrounding them. Vibrant purple phacelia creates captivating brushstrokes that paint the landscape, and lavender desert verbena lends a hint of aromatic elegance. Picture yourself speeding along a dusty road, the heat making the air shimmer, and then all of a sudden, a field of brilliant wildflowers burst into view.

The secret to this incredible show? The valley was pounded by the unusually heavy rains. Seeds, lying dormant in the dry dirt for years, sensed a change in the weather and rose to the surface when the long-awaited kiss of rain arrived. Full of new life, they pushed through the cracked dirt, reaching for the light. The result was a display of wildflowers unlike anything Death Valley had ever seen.

Only the flowers were celebrating, though, as a secret world teemed with life beneath the vivid canopy. Tiny insects fluttered from blossom to bloom, feasting on the sweet nectar, their bodies glistening like diamonds. Drawn by the promise of a bountiful meal, lizards emerged from their burrows, their scales shimmering in the light, bringing with them the soft buzz of life that had previously filled the once quiet and motionless desert.

Travelers flocked to the park, excited to see this show for themselves. The gateway town of Shoshone saw a boom that had not been seen in many years. "People really came out for the sensational and fell in love with the beauty and charm and uniqueness of the desert," said one person who had never given this dry region any thought. Pictures of the "super wildflowers," as they were called, appeared on social media in droves. Owner of the nearby ecotourism firm Susan Sorrells remarked, "We just started booming to the point we almost couldn't keep up with the business," with a hint of joy in her voice.

A delicate dance of life, every species contributing to the vast desert symphony, the desert bloom supported the entire ecosystem and was more than just an aesthetic delight. Pollinators were drawn to the vivid blossoms, ensuring the survival of plant life. The local lizards in the desert, on the other hand, enjoyed the insects as a delectable feast. The lizards were an important source of food for larger predators like coyotes.

But amid the celebrations, there was a somber reminder that, while the desert was undeniably resilient—it was truly amazing how it could adapt to such drastic changes—the record-breaking rains that had brought forth the breathtaking bloom was a warning of things to come. Climate experts predicted more extreme weather events, like longer, more severe droughts punctuated by heavy downpours. If the ecosystem's delicate balance was tipped too far, this once-in-a-lifetime display might turn into a fleeting sight, a reminder of the desert's unrealized potential.

Awe-inspiring and enlightening, the tale that seemed to have been spoken on the wind in the desert, Death Valley's desert bloom served as a reminder of nature's tenacity. It was a brilliant show of color and vitality in an area where solitude is the norm for most of the year. It served as a reminder that life finds a way and beauty can bloom in the most inhospitable places.

Chapter 4: In an Extreme Future, Can Death Valley Withstand the Whiplash?

Death Valley has always been tough. For millennia, the desert winds, the searing heat, and the unyielding sun have sculpted this brutal landscape. Despite this, life has continued to thrive here with amazing tenacity. Lately, though, things have gotten a little more fierce.

The harshness of this environment is exemplified by the bleached bones of creosote bushes, some of which are older than you can imagine. Imagine yourself driving across this wasteland, the sun shining unceasingly, the broken dirt gleaming, and then a burst of color appears around a curve in the road; fields of brilliant wildflowers cover the valley bottom, larger and more vigorous than anyone has ever seen, a strange mix of life and death that looks like something from a science fiction movie.

Scientists have long predicted a future in which deserts will be hotter and drier, but every now and then a storm will have significantly more force. This is Death Valley's new reality, my friends. an area ravaged by extreme weather, with record-breaking floods and droughts succeeding one another.

Think about it this way: remember the unexpected floods that turned Death Valley's road into a raging river? "Flashier storms" are those storms that happen more frequently because a warmer atmosphere holds on to moisture longer and releases more of it when it does. These torrential downpours may seem like a lifesaver, but they are not a sustainable solution because the dry earth cannot absorb so much water at once, leading to erosion and flash floods.

What does this mean for Death Valley's future, and will the species that inhabit this harsh environment be able to adapt to the harsh climate?

That's the fundamental question that everyone is trying to answer. The problem with living in the desert is that it's all about survival. Many plants in the desert have developed incredible mechanisms to withstand drought. Some plants, like the creosote bush, have massive root systems that delve into underground water supplies. Some plants, like some varieties of wildflowers, produce seeds that can lie dormant for hundreds of years, waiting for the right conditions to arise.

Even animals can be resourceful. The Mojave Desert tortoise, for example, can burrow underground and survive for months on fat reserves during a drought. Female tortoises can even hold onto sperm for years to wait for the right time to fertilize their eggs. "Hey, things might get tough, but we've got this!"is the way that nature speaks to us.

Nevertheless, even the toughest will have their limits when Death Valley gets hotter and drier, upsetting the delicate balance between rainfall and evaporation. Millennia-old plants might find it difficult to find enough water, and animals that have evolved to survive droughts might find themselves in more extended droughts than in the past.

It's not just about Death Valley, though; this is a story being told in deserts all over the world, and the changes we're seeing here are a taste of what's to come for other arid regions as these fragile ecosystems are at the vanguard of climate change.

But there is hope. Remember those amazing wildflowers that sprung up after the historic rainstorm? That's proof of the desert's remarkable resilience; the rain restored the ecosystem's "seed bank" and breathed new life into the parched ground. These seeds will stay dormant until they have the opportunity to bloom and bring color to the parched ground.

Death Valley's and similar desert places' fate will depend on how carefully humans and the environment coexist. To give these ecosystems a fighting chance, we need to find ways to reduce the effects of climate change and our carbon footprint.

But it's also about appreciating the tenacity and beauty of these wild yet attractive landscapes. Death Valley is notorious for its extremes, but life finds a way to survive there even in the most unlikely places. By understanding the challenges it faces, we can learn a great deal about resilience, adaptability, and the delicate balance of our planet.

www.ingramcontent.com/pod-product-compliance
Lightning Source LLC
Chambersburg PA
CBHW070957220526
45471CB00007B/3074